The Journey Beyond Earth

Inside Space Historic Mission—From the World's First Commercial Spacewalk to the Edge of Space and the Bold Future of Private Space Exploration

Saraline Tusks

Copyright © 2024 by Saraline Tusks

All rights reserved. No part of this publication may be reproduced, distributed, or transmitted in any form or by any means, including photocopying, recording, or other electronic or mechanical methods, without the prior written permission of the publisher, except in the case of brief quotations embodied in critical reviews and certain other noncommercial uses permitted by copyright law

Table of Contents

Introduction..5
    The New Frontier of Space: Why Polaris Dawn Matters........5
    SpaceX and the Evolution of Commercial Space Travel.........7
    A Glimpse into the Polaris Dawn Mission.............................9

Chapter 1: The Crew Behind the Mission..........................13
    Jared Isaacman: The Visionary Billionaire.............................13
    Scott "Kidd" Poteet: The Wingman......................................15
    Anna Menon and Sarah Gillis: Engineers and Trailblazers...17
    The Making of a Dream Team..19

Chapter 2: Preparing for the Impossible............................21
    Training for Space: From Zero Gravity to Survival Drills.....21
    Building the Crew Dragon Capsule: Engineering Marvels...24
    The Challenge of a Commercial Spacewalk.........................28
    Overcoming the Unknowns of Deep Space........................31

Chapter 3: Reaching New Heights......................................35
    Launch Day: A Momentous Leap into the Cosmos.............35
    Breaking Records: Flying Higher Than Ever Before.............38
    The Significance of 870 Miles: Outpacing Apollo's Legacy. 42
    Life in Orbit: What It Takes to Live and Work in Space........46

Chapter 4: The First Commercial Spacewalk......................51
    A Bold Step into the Void..51
    Sarah Gillis and Jared Isaacman: Pioneers in Action.............53
    Testing the EVA Suits: Ensuring Survival in the Vacuum.... 56

The Spacewalk Experience: A View Like No Other............. 59

Chapter 5: Science and Humanity Among the Stars........... 63

40 Experiments in 5 Days: Advancing Space Research.........63

Space Adaptation Syndrome: Understanding Human Physiology in Microgravity..67

Sarah Gillis' Musical Tribute: Art Meets Science in Space....70

The Power of Story: Reading "Kisses From Space" to St. Jude Patients..74

Chapter 6: Returning Home: The Journey Back to Earth. 77

The De-Orbit Burn: Preparing for Re-entry.........................77

Surviving the Heat: Protecting the Crew from 3,500°F....... 79

Splashdown in the Gulf: A Safe Return to Earth..................82

What Comes Next for SpaceX and Private Space Exploration. 84

Conclusion.. 87

Introduction

The New Frontier of Space: Why Polaris Dawn Matters

The Polaris Dawn mission represents a pivotal moment in the history of human space exploration. Not only does it push the boundaries of commercial space travel, but it also opens the door to a future where space is accessible to more than just government astronauts. As the first of the Polaris Program's three planned missions, Polaris Dawn sets a precedent for what can be achieved when private industry collaborates with space agencies to tackle challenges that were once thought to be the sole domain of national space programs.

Polaris Dawn's significance extends beyond the excitement of launching into orbit. It aims to break new ground in several areas, including conducting the first-ever commercial spacewalk, testing advanced communication technologies, and performing critical scientific experiments that could have a profound impact on the future of space exploration. These breakthroughs are not just about pushing technological limits; they have the potential to change our understanding of human biology, communication, and life in space.

For decades, space was a frontier limited to government organizations like NASA, Roscosmos, and the European Space Agency. The advent of private players, particularly SpaceX, has democratized space, creating a new frontier where private missions like Polaris Dawn can pave the way for larger and more ambitious endeavors. The mission's focus on education, research, and inclusivity — evident in its diverse crew — signals that space exploration is no longer just about national pride or scientific prestige; it's about advancing humanity as a whole.

SpaceX and the Evolution of Commercial Space Travel

The rise of SpaceX has fundamentally altered the landscape of space travel. Founded by Elon Musk in 2002, SpaceX's goal was to revolutionize space technology, with the ultimate aim of enabling humans to live on other planets. Over the years, SpaceX has introduced a series of groundbreaking innovations, including the development of the Falcon 9, the world's first reusable rocket, and the Dragon spacecraft, which has successfully transported astronauts to the International Space Station (ISS) as part of NASA's Commercial Crew Program.

What sets SpaceX apart is its ability to reduce the cost of launching into space, making space travel more affordable and thus more accessible to private entities. In the past, launching a spacecraft required the use of single-use rockets that burned millions of dollars per mission. SpaceX's reusable rockets, which can be relaunched multiple times, have dramatically cut costs, allowing for a greater frequency of launches and more ambitious missions.

Polaris Dawn is a direct beneficiary of this revolution. As a fully private mission, it relies on SpaceX's Falcon 9 and Dragon spacecraft, both of which have been tested and proven through numerous NASA and commercial missions. The mission also reflects SpaceX's commitment to pushing the boundaries of what is possible in space travel. By conducting the first commercial spacewalk and flying farther from Earth than any human since the Apollo missions, Polaris Dawn symbolizes SpaceX's philosophy of constant innovation and its vision for the future: making space exploration a routine part of human life.

Beyond the technological achievements, SpaceX's contribution to commercial space travel lies in its ability to inspire a new generation of space enthusiasts. By working with private citizens like Jared Isaacman, SpaceX has shown that space is no longer an exclusive club for government astronauts. Instead, it's becoming a more open frontier, one that will soon be accessible to scientists, researchers, and even tourists, thanks to the advances made by companies like SpaceX.

A Glimpse into the Polaris Dawn Mission

The Polaris Dawn mission is poised to be one of the most exciting and ambitious space endeavors in recent memory. Spearheaded by billionaire entrepreneur and space enthusiast Jared Isaacman, the mission marks the first of three planned spaceflights under the Polaris Program, which aims to push the boundaries of what private space travel can achieve. Scheduled to launch aboard SpaceX's Crew Dragon spacecraft, Polaris Dawn will carry a crew of four into space for a five-day journey designed to break records and make history.

One of the mission's most notable objectives is to perform the first-ever commercial spacewalk. While spacewalks have been a routine part of space exploration for government astronauts, Polaris Dawn's commercial spacewalk represents a significant leap forward for private space missions. The crew will don specially designed spacesuits to venture outside the spacecraft, testing SpaceX's new extravehicular activity (EVA) suits, which are critical for future missions to the Moon, Mars, and beyond. This spacewalk is not just a symbolic gesture; it serves as a testbed for future missions where private astronauts may need to perform maintenance or scientific experiments outside their spacecraft.

Polaris Dawn is also focused on scientific research. During their time in orbit, the crew will conduct around 40 different experiments aimed at advancing our understanding of human physiology in space, the impact of space travel on mental health, and the potential for medical breakthroughs that could benefit people on Earth. This research is crucial for the future of long-duration spaceflights, particularly missions to Mars, where astronauts will face unprecedented physical and psychological challenges.

Another groundbreaking aspect of the mission is its attempt to reach the highest Earth orbit ever achieved since the Apollo missions. The spacecraft is expected to travel as far as 870 miles (1,400 kilometers) above the Earth's surface, surpassing the orbits of both the ISS and the Hubble Space Telescope. This will allow the crew to test advanced communication technologies, including Starlink laser-based systems, which could revolutionize internet connectivity for future space missions and even remote areas on Earth.

Ultimately, Polaris Dawn is not just about pushing the limits of human spaceflight; it's about inspiring the next generation of explorers. The mission is closely tied to charitable causes, including raising awareness and funds for St. Jude Children's Research Hospital. Through its blend of technological innovation, scientific research, and humanitarian efforts, Polaris Dawn is much more than a space mission—it's a glimpse into the future of space exploration and the potential for private missions to advance human knowledge and capabilities.

Chapter 1: The Crew Behind the Mission

Jared Isaacman: The Visionary Billionaire

Jared Isaacman, the driving force behind the Polaris Dawn mission, is no ordinary billionaire. A seasoned entrepreneur, philanthropist, and accomplished pilot, Isaacman embodies a rare combination of business acumen, daring, and vision. His passion for space exploration has led him to fund and command not only the Polaris Dawn mission but also the groundbreaking Inspiration4 mission, which made history in 2021 as the first all-civilian crewed spaceflight. As the CEO and founder of Shift4 Payments, a payment processing company he started as a teenager, Isaacman has always had a knack for innovation and pushing boundaries. His success in the business world has given him the financial means to realize his dreams of advancing private space exploration.

Isaacman's fascination with aviation and space began at an early age. He became a licensed pilot in his teens and went on to set several world records in aviation. His background in high-speed jet flight, including flying in air shows and co-founding Draken International, a defense contractor that provides fighter jets for military training, has equipped him with the skills and mindset required for space travel. But Isaacman's aspirations go beyond personal achievement. His aim is to pave the way for future space explorers by demonstrating that private missions can push the boundaries of human spaceflight, conducting important scientific research, and advancing space technologies.

Polaris Dawn is the next step in Isaacman's mission to democratize space travel and make it more accessible to a wider audience. He views space exploration not only as an exciting frontier but also as a means to contribute to humanity's long-term survival. Isaacman has often spoken about the importance of making humanity a multi-planetary species, a goal shared by SpaceX founder Elon Musk. Isaacman's leadership and vision have been instrumental in shaping the Polaris Dawn mission, which aims to go further into space than any human has gone since the Apollo missions. Through his financial commitment and personal involvement as the mission's commander, Isaacman is laying the groundwork for the future of commercial space exploration.

Scott "Kidd" Poteet: The Wingman

Scott "Kidd" Poteet is not just Jared Isaacman's friend and wingman; he is a vital part of the Polaris Dawn crew, bringing a wealth of experience in military aviation and leadership to the mission. A retired U.S. Air Force Lieutenant Colonel, Poteet has had a distinguished career flying F-16 fighter jets and serving in various command positions. His background in high-pressure environments and his ability to make split-second decisions make him an ideal fit for space missions, where precision and teamwork are paramount.

Poteet's relationship with Isaacman goes back several years, with the two bonding over their shared love of aviation and adventure. Poteet played a significant role in the success of the Inspiration4 mission, serving as mission director and overseeing the crew's training and preparations. His leadership and operational expertise ensured that the mission ran smoothly, making him a natural choice for the Polaris Dawn mission.

As a crew member, Poteet will be responsible for supporting the mission's objectives, including the first commercial spacewalk and scientific experiments conducted during the mission. His extensive experience in flight operations, coupled with his close relationship with Isaacman, makes him a trusted and reliable partner in the cockpit. Poteet's calm demeanor and tactical mindset are key assets in an environment where every decision counts.

Anna Menon and Sarah Gillis: Engineers and Trailblazers

Anna Menon and Sarah Gillis, the two engineers on the Polaris Dawn crew, represent the next generation of space trailblazers. Both women work for SpaceX and have played pivotal roles in the company's human spaceflight programs. Their inclusion on the Polaris Dawn mission underscores the importance of diversity and technical expertise in modern space exploration.

Anna Menon, a Lead Space Operations Engineer at SpaceX, has extensive experience in mission control and crew operations. She has worked closely with NASA astronauts and played a key role in the launch of the Crew Dragon missions. Menon's background in biomedical engineering also makes her a valuable asset for the scientific experiments that will be conducted during the mission. She will be responsible for monitoring crew health and conducting research related to human physiology in space, an area of critical importance as missions grow longer and more complex.

Sarah Gillis, SpaceX's Lead Space Operations Engineer for astronaut training, has been instrumental in preparing crews for missions aboard the Crew Dragon spacecraft. She has trained NASA astronauts for missions to the International Space Station and has a deep understanding of the systems and operations that will be critical to the success of Polaris Dawn. Gillis will be responsible for ensuring the crew is fully prepared for the mission's technical and operational challenges, including the spacewalk and other complex tasks.

Both Menon and Gillis are breaking barriers as women in space exploration, a field that has historically been dominated by men. Their inclusion on the Polaris Dawn mission reflects a broader trend toward greater diversity and inclusion in space, and their expertise will be critical to the mission's success. Their presence also serves as an inspiration to young women and girls who aspire to careers in science, technology, engineering, and mathematics (STEM).

The Making of a Dream Team

The Polaris Dawn crew is more than just a group of individuals; they represent the ideal combination of vision, expertise, and teamwork required for a mission of this magnitude. Each member brings a unique set of skills and experiences to the table, creating a dynamic team capable of handling the complexities of space travel. Jared Isaacman's leadership and vision are complemented by Scott Poteet's operational expertise, while Anna Menon and Sarah Gillis provide the technical know-how that is essential for the mission's success.

What makes this crew stand out is their shared commitment to pushing the boundaries of space exploration and making a positive impact on Earth. Whether it's through advancing scientific research, testing new technologies, or raising funds for St. Jude Children's Research Hospital, each crew member is focused on contributing to something larger than themselves. Their teamwork and shared sense of purpose are what will make Polaris Dawn not just a mission, but a milestone in the ongoing journey to make space accessible to all.

As the crew prepares for their journey, they are not just training for the physical and technical challenges of space travel; they are also preparing to inspire the world. The Polaris Dawn mission is as much about exploration as it is about education and inspiration. Through their efforts, this dream team is showing the world that space is not just the final frontier, but the next great opportunity for human advancement.

Chapter 2: Preparing for the Impossible

Training for Space: From Zero Gravity to Survival Drills

Preparing for a mission like Polaris Dawn is no small feat. Astronauts must undergo rigorous training to ensure they can handle the unique and often extreme challenges of space. For the Polaris Dawn crew, this training includes everything from adapting to zero gravity to mastering survival drills for emergency situations. Each aspect of their preparation is meticulously planned to ensure they are physically and mentally ready for the mission's demanding objectives, including the first commercial spacewalk.

The first step in their training involves adapting to the absence of gravity, or microgravity, in space. This is a significant challenge because the human body is naturally accustomed to functioning in Earth's gravity. To simulate the experience of zero gravity, the crew undergoes parabolic flights, often referred to as the "vomit comet." During these flights, the aircraft performs a series of steep climbs and descents, creating brief periods of weightlessness. These training flights help the astronauts get used to moving, working, and even eating in a weightless environment, which is crucial for their comfort and efficiency during the mission.

Another essential part of their training is learning how to deal with emergency situations in space. Spacecraft malfunctions, medical emergencies, or other unforeseen events can pose significant risks during a mission, and the crew must be prepared to respond effectively. They undergo survival drills that cover a range of potential scenarios, from rapid spacecraft depressurization to emergency landings in harsh environments on Earth. For instance, the crew is trained to survive in extreme conditions such as cold water or harsh terrain, simulating what could happen if the Crew Dragon capsule reenters Earth's atmosphere and lands in an unexpected location.

Astronauts must also be in peak physical condition to endure the rigors of space travel. Space takes a toll on the human body, including muscle atrophy and bone density loss due to prolonged exposure to microgravity. As part of their preparation, the crew undergoes intense physical conditioning to strengthen their muscles, improve cardiovascular health, and enhance overall endurance. This ensures that they are not only fit to handle the physical challenges of space but also able to perform complex tasks during the mission, such as the spacewalk.

The mental aspect of space training is just as crucial as the physical. Astronauts must be able to make quick, high-stakes decisions in stressful situations, all while working as part of a tightly knit team. Crew members undergo psychological evaluations and stress management training to ensure they are mentally resilient and capable of maintaining focus and calm under pressure. This mental toughness is particularly important for Polaris Dawn, given the unprecedented nature of the mission and its potential challenges.

In the end, the extensive training for Polaris Dawn is designed to prepare the crew for every possible scenario, whether it involves floating in zero gravity or surviving in the wilderness after an emergency landing. The goal is not only to ensure the safety of the crew but also to give them the confidence and skills needed to successfully complete the mission's ambitious objectives.

Building the Crew Dragon Capsule: Engineering Marvels

The Polaris Dawn mission will be carried out aboard SpaceX's Crew Dragon capsule, an engineering marvel that has revolutionized commercial space travel. The Crew Dragon was designed with both safety and efficiency in mind, making it one of the most advanced spacecraft ever built. For Polaris Dawn, the spacecraft has been further refined and customized to meet the specific requirements of the mission, including the ability to conduct the first commercial spacewalk.

One of the key features of the Crew Dragon is its automated flight system, which allows the spacecraft to autonomously dock with the International Space Station (ISS) and other space vehicles. This system reduces the need for manual piloting, although the crew is trained to take control in the event of a system failure. The spacecraft is also equipped with advanced navigation and communication systems, which ensure that the crew remains in constant contact with mission control on Earth.

The interior of the Crew Dragon is designed to maximize comfort and functionality. The capsule can carry up to seven astronauts, but for Polaris Dawn, it will accommodate a crew of four. The seats are made from carbon fiber and are custom-fitted for each astronaut, ensuring maximum comfort during the launch, orbit, and reentry phases of the mission. The capsule is also equipped with touchscreen controls, allowing the crew to monitor and manage various systems with ease.

Safety is paramount in the design of the Crew Dragon, and the spacecraft is equipped with several systems to ensure the well-being of the crew. One of the most important safety features is the launch escape system, which is designed to quickly propel the capsule away from the rocket in the event of an emergency during launch. This system uses a set of SuperDraco thrusters that can fire in a fraction of a second, ensuring the crew's safety even in the most critical situations.

In addition to its safety features, the Crew Dragon is also designed to support scientific research. The spacecraft is equipped with storage compartments for scientific instruments and experiments, making it a valuable platform for conducting research in microgravity. For Polaris Dawn, the Crew Dragon will carry out several scientific experiments aimed at advancing our understanding of space and improving technologies for future missions.

The spacewalk planned for Polaris Dawn presents unique engineering challenges, and the Crew Dragon has been modified to support this milestone event. SpaceX engineers have developed custom space suits and airlock systems to enable the crew to safely exit the spacecraft and conduct the spacewalk. This marks a significant advancement in the design of the Crew Dragon, as it pushes the boundaries of what is possible for commercial space exploration.

The Challenge of a Commercial Spacewalk

One of the most ambitious objectives of the Polaris Dawn mission is the first-ever commercial spacewalk. Spacewalks, or extravehicular activities (EVAs), are among the most challenging and dangerous aspects of space travel. They require astronauts to leave the safety of their spacecraft and venture into the vacuum of space, where they are exposed to extreme temperatures, radiation, and the constant threat of micrometeoroid impacts. For a commercial crew to undertake this task is a groundbreaking step in the evolution of space exploration.

A successful spacewalk requires a combination of advanced technology, extensive training, and meticulous planning. For Polaris Dawn, the crew will don custom-built space suits designed by SpaceX engineers specifically for this mission. These suits must protect the astronauts from the harsh environment of space while providing them with the flexibility and dexterity needed to carry out tasks outside the spacecraft. The suits are equipped with life support systems, including oxygen tanks, carbon dioxide scrubbers, and temperature regulation systems, to ensure that the astronauts can survive for extended periods outside the spacecraft.

The spacewalk also presents logistical challenges. The crew must carefully plan their movements to avoid drifting away from the spacecraft. They will be tethered to the Crew Dragon using a combination of mechanical tethers and jetpacks that allow for precise movement in space. The crew will also need to maintain constant communication with mission control to ensure their safety and coordination during the spacewalk.

One of the primary goals of the Polaris Dawn spacewalk is to test new technologies that will be essential for future missions to the Moon, Mars, and beyond. The crew will carry out experiments designed to gather data on how the human body responds to the conditions of space, particularly the effects of radiation and microgravity. This data will be invaluable for planning longer-duration space missions and ensuring the safety of future astronauts.

The commercial spacewalk is a significant milestone for both SpaceX and the broader space industry. It demonstrates that private companies can not only send humans into space but also undertake complex and dangerous tasks traditionally reserved for government space agencies like NASA. By successfully completing the spacewalk, Polaris Dawn will help pave the way for future commercial space missions that push the boundaries of human exploration.

Overcoming the Unknowns of Deep Space

Space exploration is fraught with unknowns, particularly when venturing into the deeper regions of space. Unlike missions to low Earth orbit, where the International Space Station orbits, deep space missions expose astronauts to greater risks, including higher levels of radiation, communication delays, and the psychological challenges of being farther from Earth. For Polaris Dawn, overcoming these unknowns is a critical part of the mission's objectives, as it seeks to expand humanity's understanding of space and prepare for future missions beyond Earth's orbit.

One of the most significant challenges of deep space exploration is radiation. The further astronauts venture from Earth, the less protection they have from the planet's magnetic field, which shields them from cosmic rays and solar radiation. Prolonged exposure to these high-energy particles can have severe health effects, including an increased risk of cancer and damage to the central nervous system. To mitigate these risks, the Polaris Dawn crew will rely on advanced shielding technologies built into the Crew Dragon capsule and their space suits. Additionally, part of the mission's research will focus on gathering data about the levels of radiation encountered during the mission, which will be used to improve safety measures for future deep space missions.

Another challenge is communication. As the crew travels farther from Earth, the time delay in sending and receiving messages increases. In low Earth orbit, communication with mission control is almost instantaneous, but in deep space, even a small delay can have significant consequences. This delay requires astronauts to be more self-reliant, as they may need to make critical decisions without immediate input from mission control. To prepare for this, the Polaris Dawn crew has undergone extensive training in autonomous decision-making and problem-solving, ensuring that they can handle unexpected situations without outside assistance.

The psychological challenges of deep space are also considerable. Astronauts in deep space missions are isolated from the familiar sights and sounds of Earth, and the vast emptiness of space can be mentally taxing. Polaris Dawn's crew will be exposed to these conditions for an extended period, providing valuable insights into how humans can cope with the psychological demands of deep space travel. The mission's findings will inform future efforts to prepare astronauts for the even longer and more isolated missions required for journeys to Mars and other distant destinations.

Despite the many challenges, overcoming the unknowns of deep space is essential for the future of space exploration

. The Polaris Dawn mission will not only push the boundaries of what is possible for commercial spaceflight but also contribute valuable knowledge that will be crucial for humanity's efforts to explore and eventually colonize other worlds. By tackling the unknowns head-on, the Polaris Dawn crew is helping to pave the way for a future where space travel is no longer the domain of a select few but an achievable goal for all.

Chapter 3: Reaching New Heights

Launch Day: A Momentous Leap into the Cosmos

Launch day for the Polaris Dawn mission is more than just another rocket firing into the skies. It represents a momentous leap into the cosmos, marking a new chapter in the evolution of human space exploration. The culmination of years of planning, engineering, and rigorous training, this day is filled with excitement, anxiety, and hope as the crew prepares to leave Earth and venture into the unknown.

As the countdown begins, the crew, clad in their SpaceX-engineered flight suits, will sit inside the Crew Dragon capsule atop a Falcon 9 rocket, their senses heightened by the magnitude of the moment. The Falcon 9, a two-stage reusable rocket developed by SpaceX, will be responsible for propelling the astronauts into space. It has earned a reputation as a reliable workhorse, having previously carried NASA astronauts, commercial payloads, and international partners to the International Space Station (ISS) and beyond.

For the Polaris Dawn crew, the moments leading up to liftoff will be a blur of final system checks and communication with mission control. Once all systems are "go," the rocket engines will ignite, and the tremendous force of the Falcon 9's 1.7 million pounds of thrust will push the Crew Dragon off the launchpad. Within seconds, the crew will feel the sheer power of the rocket accelerating them toward the heavens, their journey to the edge of space finally underway.

The launch itself is both a physical and psychological test. Astronauts experience several times the force of gravity, or g-forces, as the rocket ascends through the atmosphere, compressing them into their seats. Meanwhile, their training and mental fortitude are put to the test as they monitor the spacecraft's systems, communicating with mission control while adapting to the intense vibrations and forces of the launch.

For those watching from the ground, launch day is a symbol of human ambition and technological prowess. Millions of viewers around the world will tune in to witness this historic event, many of them awed by the sheer spectacle of a rocket breaking free from Earth's gravity. But beyond the awe and excitement, launch day carries profound significance. It represents humanity's unyielding desire to explore, to push boundaries, and to understand what lies beyond our home planet.

Polaris Dawn is not just another space mission; it is part of a broader vision to make space more accessible to ordinary people. With a commercial crew of private citizens, this launch day signifies a turning point in space travel. No longer is space exploration solely the domain of government space agencies; it is increasingly becoming a frontier where private enterprises and individuals can make their mark.

As the Falcon 9 soars higher and higher, eventually reaching the vacuum of space, the booster stage will detach and return to Earth, landing on a drone ship for future reuse. The Crew Dragon capsule will then continue its journey, preparing to reach new heights that no human has ever before achieved in a commercial mission. For the crew of Polaris Dawn, launch day will mark the start of an adventure that transcends Earth, pushing the boundaries of what is possible for human space exploration.

Breaking Records: Flying Higher Than Ever Before

The Polaris Dawn mission is set to make history by breaking new records and pushing the limits of human space exploration. One of the most remarkable aspects of this mission is that it will fly higher than any previous commercial space mission, reaching an altitude of approximately 870 miles above Earth. This unprecedented achievement not only sets new benchmarks for commercial spaceflight but also brings humanity closer to realizing the long-term goal of deep space exploration.

Historically, the altitude of manned space missions has been limited to low Earth orbit (LEO), which extends up to about 1,200 miles above the Earth. The majority of human spaceflight, including missions to the International Space Station (ISS), takes place within this range. By venturing beyond LEO, Polaris Dawn is breaking through a barrier that has constrained human space exploration for decades. The mission will take the crew farther than any commercial astronauts have ever gone, surpassing previous milestones set by SpaceX's own Crew Dragon flights.

This mission holds special significance because of its commercial nature. While governmental space agencies like NASA have long held the record for human spaceflight, Polaris Dawn's success would mark a turning point where private companies take the lead in pushing the envelope of space travel. This represents a shift toward the democratization of space, where the potential for exploration is no longer exclusive to governments with vast resources.

Flying higher means facing new challenges, and the crew of Polaris Dawn will be tasked with overcoming a host of technical and environmental obstacles. The higher the altitude, the greater the exposure to cosmic radiation and the harsher the conditions of space. Beyond the protective layers of Earth's atmosphere, the crew will be more vulnerable to solar storms and cosmic rays, which pose significant risks to human health and spacecraft systems. To mitigate these risks, the Crew Dragon capsule has been outfitted with advanced radiation shielding and other protective measures, ensuring the crew's safety during their record-breaking journey.

In addition to the heightened risks, the crew will also have the opportunity to observe and study aspects of space that are not possible at lower altitudes. From 870 miles above Earth, they will be able to capture images and collect data that provide insights into our planet's atmosphere, magnetic field, and the near-Earth environment. This research will be invaluable for future space missions, especially those that aim to venture even farther, such as planned missions to Mars or the Moon.

By breaking records and flying higher than ever before, Polaris Dawn is setting the stage for the next generation of space exploration. The mission's success will not only inspire future space travelers but also help pave the way for deeper space exploration, bringing humanity closer to realizing its dream of reaching other planets. For SpaceX and the Polaris Dawn crew, this journey represents the culmination of years of technological advancements and bold vision. The mission stands as a testament to human ingenuity and the drive to explore the cosmos.

The Significance of 870 Miles: Outpacing Apollo's Legacy

The Polaris Dawn mission is set to achieve an altitude of 870 miles above Earth—an unprecedented feat for a commercial space mission. This altitude surpasses that of any human spaceflight in decades, except for those of the Apollo missions, which famously took astronauts to the Moon. While Apollo's legacy is monumental, Polaris Dawn is blazing a new trail by reaching a similar height for an entirely different purpose. In doing so, it is redefining the significance of human space travel and exploring new frontiers in space exploration.

For context, the Apollo missions, particularly Apollo 11, flew to the Moon at an altitude of approximately 238,000 miles, marking humanity's first steps on another celestial body. However, after the last Apollo mission in 1972, manned space missions have been confined to low Earth orbit (LEO), with the primary focus on the International Space Station (ISS) and satellite deployment. Polaris Dawn, by venturing far beyond LEO to an altitude of 870 miles, is charting a new course for future space exploration, opening the door for more ambitious endeavors in space.

The altitude of 870 miles is significant because it represents a transition between LEO and the more distant regions of space known as medium Earth orbit (MEO) and beyond. At this altitude, the spacecraft will experience conditions that are drastically different from those in LEO. The Earth's atmosphere will be much thinner, and the crew will be exposed to increased levels of cosmic radiation.

These factors make Polaris Dawn not only a pioneering mission for commercial spaceflight but also a crucial experiment in understanding the human body's ability to withstand the hazards of space at higher altitudes.

Outpacing Apollo's legacy in terms of altitude is not the sole aim of Polaris Dawn, but it adds symbolic weight to the mission. It serves as a reminder that space exploration did not end with Apollo and that humanity's drive to explore continues to thrive. While the Apollo program was about reaching the Moon, Polaris Dawn is about pushing the boundaries of what is possible for commercial spaceflight, setting the stage for future missions to distant destinations like Mars or asteroids. It's a bridge between the past and the future, leveraging the knowledge gained from Apollo while pursuing a new era of space travel.

Reaching 870 miles also offers scientific opportunities that were not fully explored during the Apollo era. From this vantage point, the crew will be able to study Earth's atmosphere and magnetic field in ways that could enhance our understanding of climate change, satellite technology, and space weather. Moreover, the mission will gather data that will inform future deep-space explorations, such as how prolonged exposure to cosmic radiation affects the human body and spacecraft systems.

Ultimately, the significance of Polaris Dawn's 870-mile altitude lies not only in the breaking of records but in what it represents for the future of space exploration. It is a bold step toward a future where space is accessible to more people, where commercial missions push the boundaries once reserved for government agencies, and where the exploration of deeper space becomes a tangible reality. By outpacing Apollo's legacy in terms of altitude, Polaris Dawn is signaling that the next great age of space exploration is just beginning.

Life in Orbit: What It Takes to Live and Work in Space

Living and working in space is a challenge unlike any other, and for the crew of Polaris Dawn, adapting to life in orbit will be a crucial aspect of their mission. While the glamour of space travel often captures the public's imagination, the reality is that astronauts face numerous physical, mental, and logistical challenges during their time away from Earth. Whether it's the effects of microgravity on the body, the psychological strain of isolation, or the complexities of carrying out tasks in a weightless environment, life in orbit demands a high level of preparation and adaptability.

One of the most immediate challenges of living in space is the effect of microgravity on the human body. On Earth, gravity constantly pulls down on our bodies, but in space, astronauts experience weightlessness. This might seem fun at first—floating around the spacecraft and performing somersaults in mid-air—but over time, the lack of gravity takes a toll on the body. Muscles begin to atrophy, bones lose density, and fluids shift to the upper body, causing facial puffiness and pressure on the eyes. To mitigate these effects, astronauts must engage in rigorous exercise routines, often spending up to two hours a day using specialized equipment like treadmills and resistance machines designed for use in microgravity.

Beyond the physical challenges, there are also psychological factors to consider. Space can be a lonely and isolating place. Although the Polaris Dawn crew will have each other for company, they will be far from home, family, and friends. The psychological strain of being in such an unfamiliar and confined environment can lead to stress, anxiety, and even depression. To combat this, space agencies and mission planners place a strong emphasis on mental health, providing astronauts with regular communication with loved ones, access to psychological support, and carefully planned schedules to keep them mentally stimulated and engaged.

Work in space is another complex undertaking. Even simple tasks, such as eating or sleeping, become more challenging in a weightless environment. Astronauts have to secure themselves to a sleeping area to avoid floating around, and meals are carefully prepared to minimize crumbs and liquids, which can be hazardous in a spacecraft. Every action must be carefully planned and executed, with astronauts needing to master the art of maneuvering in microgravity, often using handholds or foot restraints to anchor themselves while performing tasks.

For the Polaris Dawn crew, the mission will also involve performing scientific experiments and, potentially, the first commercial spacewalk. Working in space requires intense concentration, precision, and coordination. During a spacewalk, astronauts must wear bulky spacesuits and work outside the spacecraft, exposed to the harsh conditions of space. They must be prepared to navigate in a weightless environment while dealing with extreme temperatures, radiation, and the constant threat of debris. The crew's ability to function effectively in these conditions will be critical to the mission's success.

Despite the challenges, life in orbit offers unique rewards. Astronauts on Polaris Dawn will have the rare privilege of seeing Earth from space, experiencing the breathtaking view of the planet from hundreds of miles above. This perspective, known as the "overview effect," has been described by many astronauts as life-changing, instilling a profound sense of interconnectedness and a desire to protect our planet.

For the crew of Polaris Dawn, life in orbit will be both challenging and awe-inspiring. Their ability to adapt, work, and thrive in the harsh environment of space will be a testament to human resilience and ingenuity, pushing the boundaries of what is possible and setting the stage for the future of space exploration.

Chapter 4: The First Commercial Spacewalk

A Bold Step into the Void

The first commercial spacewalk is set to be a groundbreaking moment in the history of human space exploration, as it will mark the first time a crew of private astronauts steps out into the vacuum of space. This bold venture, a key component of the Polaris Dawn mission, will pave the way for future commercial space operations and demonstrates the rapid advancements being made in private space travel.

Historically, extravehicular activities (EVAs), or spacewalks, have been the domain of government-backed astronauts from space agencies such as NASA, Roscosmos, and the European Space Agency. The first spacewalk was conducted by Russian cosmonaut Alexei Leonov in 1965, followed closely by American astronaut Ed White. Since then, EVAs have become routine but are still extremely dangerous and require significant preparation. Until now, no private citizens have attempted a spacewalk, making the Polaris Dawn mission a bold leap forward.

Stepping out into the void of space is not just a technical challenge; it is a psychological one as well. Spacewalks involve leaving the relative safety of the spacecraft and venturing into an environment that is hostile to human life. Astronauts must contend with extreme temperatures, radiation, and the constant threat of micrometeoroids or space debris, any of which could cause catastrophic damage to their suit. The space around them is completely silent, with no atmosphere to carry sound, and their only connection to the spacecraft is a tether or jet propulsion system.

This first commercial spacewalk is significant not only for the technical achievements it represents but also for the new possibilities it opens for private space exploration. By proving that private astronauts can safely perform spacewalks, the Polaris Dawn mission is pushing the boundaries of what commercial space travel can accomplish. Future missions to the Moon, Mars, or other celestial bodies will likely involve multiple EVAs, and this mission serves as a testbed for the systems, suits, and techniques needed to make those missions a reality.

As they step out into the void, the astronauts on Polaris Dawn will be pioneers, taking a bold and historic step that could change the future of human spaceflight forever.

Sarah Gillis and Jared Isaacman: Pioneers in Action

Two key figures in the upcoming first commercial spacewalk are Sarah Gillis and Jared Isaacman, both of whom are set to play pivotal roles in the

Polaris Dawn mission. Their involvement in this groundbreaking event highlights their dedication, skill, and bravery, as they prepare to become the first private citizens to embark on a spacewalk, setting new precedents for space exploration.

Jared Isaacman, a billionaire entrepreneur, is no stranger to pushing boundaries. As the commander of the Polaris Dawn mission, Isaacman's leadership is crucial to the success of this commercial endeavor. Having previously flown on SpaceX's Inspiration4 mission, Isaacman already has experience as a private astronaut, which makes him a strong candidate for such an ambitious task. However, a spacewalk is far more complex and dangerous than the missions he has previously undertaken, adding a new layer of responsibility to his role. As the first commercial astronaut to walk in space, Isaacman's place in history is guaranteed.

Sarah Gillis, the lead space operations engineer at SpaceX, will also be a vital participant in this spacewalk. Her role on the mission is particularly significant, as she represents the growing role of women in space exploration. While spacewalks have historically been male-dominated, Gillis' participation marks a step toward greater gender diversity in this challenging and prestigious field. With her background in training astronauts and designing operational systems, Gillis brings critical expertise that will ensure the spacewalk's success.

The pairing of Isaacman and Gillis as pioneers in this venture demonstrates how private spaceflight is evolving to include not only highly trained engineers but also passionate leaders who bring their own vision to space exploration. Both Isaacman and Gillis have undergone extensive training, including simulations of the spacewalk, testing the spacesuits, and practicing the techniques they will use during the EVA. Their combined skills, leadership, and courage will be tested to the limit as they take on one of the most dangerous tasks in human spaceflight.

As pioneers in action, Isaacman and Gillis are proving that space is no longer the exclusive domain of government-backed astronauts. Their participation in the first commercial spacewalk signals a new era, where private citizens can contribute to the bold pursuit of space exploration and open new pathways for humanity to follow.

Testing the EVA Suits: Ensuring Survival in the Vacuum

One of the most critical aspects of preparing for the first commercial spacewalk is the rigorous testing of the extravehicular activity (EVA) suits. These specialized suits are essentially mini spacecraft designed to protect astronauts from the harsh environment of space. Ensuring their reliability is crucial for the success of any spacewalk, as a single malfunction could mean the difference between life and death.

SpaceX, which is leading the Polaris Dawn mission, has designed custom EVA suits for the mission. These suits are an evolution of the SpaceX flight suits that have already been used in the Crew Dragon missions, but with significant modifications to enable the astronauts to survive and work outside the spacecraft. The suits must protect the crew from the vacuum of space, extreme temperatures, and high levels of radiation, all while providing enough flexibility and dexterity to perform tasks during the EVA.

Before the spacewalk, these EVA suits undergo extensive testing to ensure they can withstand the harsh conditions of space. This testing includes vacuum chamber tests, where the suits are subjected to conditions that mimic the vacuum of space. Engineers check for leaks, temperature control, and the integrity of the life support systems, which provide astronauts with oxygen and remove carbon dioxide. Additionally, the suits must be capable of withstanding micrometeoroids and space debris, which can travel at speeds of up to 17,500 miles per hour in orbit.

The Polaris Dawn crew, including Sarah Gillis and Jared Isaacman, will spend hours practicing how to move and work in the suits under weightless conditions, both in water tanks and aboard parabolic flights that simulate the microgravity environment of space. These training exercises help the crew become accustomed to the suit's limitations, such as reduced mobility and visibility, while also teaching them how to perform tasks efficiently in space. They will practice securing themselves to the spacecraft, using tools, and communicating with the rest of the crew and mission control.

One of the unique challenges of this mission is that it will involve the first use of commercial EVA suits, rather than those designed and tested over decades by NASA. While SpaceX has proven itself capable of creating reliable spacecraft, this spacewalk represents a new frontier. The success of the EVA suits will not only ensure the crew's safety but also set the standard for future commercial spacewalks.

Ultimately, the testing and validation of the EVA suits are vital to the mission's success. These suits are the crew's only barrier between survival and the harsh vacuum of space, and their performance during the spacewalk will be a testament to the advancements made in commercial spaceflight.

The Spacewalk Experience: A View Like No Other

For astronauts, a spacewalk is one of the most exhilarating and awe-inspiring experiences of their careers. The opportunity to step outside the confines of a spacecraft and float freely in the vacuum of space offers a perspective on the universe that is truly unparalleled. For the crew of Polaris Dawn, their upcoming commercial spacewalk will provide them with a view like no other—one that few humans have ever had the privilege of witnessing.

During a spacewalk, astronauts are tethered to the spacecraft to prevent them from floating away into space. Despite this safety precaution, the sensation of floating above Earth, with no physical boundaries or constraints, is described as both thrilling and humbling. From this vantage point, astronauts can see the curvature of the Earth, the thin blue line of the atmosphere, and the vast expanse of the cosmos stretching out in all directions. The beauty of the planet from space is often said to evoke a profound sense of awe, as well as a deep appreciation for the fragility and interconnectedness of life on Earth.

For the Polaris Dawn crew, this view will be particularly special. As the first private citizens to experience a spacewalk, they will be making history while taking in the breathtaking sights of our planet from hundreds of miles above. The sensation of weightlessness, combined with the visual spectacle of Earth from space, is often described as transformative, leading many astronauts to develop a new outlook on life and humanity's place in the universe—a phenomenon known as the "overview effect."

However, the spacewalk experience is not just about taking in the view. Astronauts must remain focused on their tasks, which often involve performing maintenance, repairs, or scientific experiments outside the spacecraft. The crew of Polaris Dawn will likely have a specific set of objectives to accomplish during their EVA, such as testing new tools, collecting data, or assessing the performance of their EVA suits in the vacuum of space.

Communication with mission control is critical during a spacewalk, as the astronauts must rely on their team on the ground to monitor their progress and provide guidance if any issues arise. The crew's safety is paramount, and every step of the spacewalk is carefully choreographed and rehearsed to minimize risks.

Despite the challenges, the opportunity to step into the void of space and experience the universe in such an intimate way is something that very few people ever get to experience. For Sarah Gillis, Jared Isaacman, and the rest of the Polaris Dawn crew, the first commercial spacewalk will be an unforgettable experience, offering them a view like no other and a moment in history that will inspire future generations of space explorers.

Chapter 5: Science and Humanity Among the Stars

40 Experiments in 5 Days: Advancing Space Research

The Polaris Dawn mission stands as a landmark in human space exploration, not just because of its audacious spacewalk and commercial nature, but also for its ambitious scientific agenda. Within the span of five days, the crew plans to conduct over 40 experiments, each aimed at advancing space research in ways that have the potential to benefit humanity both in space and on Earth. This accelerated pace of research represents the growing role of private space missions in contributing to the global understanding of space science, which has historically been driven by government agencies like NASA and Roscosmos.

One of the most significant aspects of Polaris Dawn's scientific mission is its focus on understanding how the human body responds to the extreme environment of space. Microgravity, intense radiation, and isolation are just a few of the factors that affect astronauts during their time in orbit. The mission's experiments will provide invaluable data on how these conditions impact human physiology, with a particular emphasis on cardiovascular health, immune function, and bone density.

Beyond human physiology, the experiments will also focus on advancing space technology. For instance, the mission will test the capabilities of new communication systems that could enable more effective interactions between astronauts and mission control during long-duration spaceflights. The mission will also examine ways to improve the effectiveness of life support systems, which are crucial for sustaining human life in deep space. This research is essential for future missions to the

Moon, Mars, and beyond, where astronauts will need to operate independently of Earth for extended periods.

Another area of focus is materials science, with experiments designed to test how various materials respond to the harsh conditions of space. These findings could lead to the development of more durable spacecraft, habitats, and equipment, ensuring that future space missions are safer and more sustainable. In addition to their space applications, many of these advancements in materials science have the potential to benefit industries on Earth, from construction to electronics.

The Polaris Dawn crew will also engage in experiments that look beyond the purely scientific, exploring how humans can thrive, not just survive, in space. This includes research on mental health, stress management, and teamwork in extreme environments. As humans move closer to becoming a multi-planetary species, these questions become increasingly important. How do we build communities in space? How can we ensure the psychological well-being of astronauts during long-duration missions? The answers to these questions will shape the future of space exploration, and the Polaris Dawn mission is at the forefront of this research.

Ultimately, the 40 experiments conducted during Polaris Dawn are not just about advancing human knowledge; they are about preparing humanity for the next phase of space exploration. By pushing the boundaries of what is possible in space research, this mission is paving the way for future generations of astronauts, scientists, and explorers. The findings from these experiments will have lasting implications, contributing to a better understanding of space and laying the foundation for future missions that will take humanity deeper into the cosmos.

Space Adaptation Syndrome: Understanding Human Physiology in Microgravity

Space adaptation syndrome (SAS), often referred to as space motion sickness, is one of the key physiological challenges that astronauts face when they first enter the microgravity environment of space. As humans are adapted to life on Earth with its constant gravitational pull, the transition to a weightless environment can cause disorientation and discomfort. During the Polaris Dawn mission, the crew will experience this firsthand, providing an opportunity for researchers to study SAS more thoroughly and potentially develop strategies to mitigate its effects.

SAS is caused by the mismatch between the signals sent from the inner ear, which helps control balance, and the lack of gravity experienced in space. On Earth, the inner ear detects the pull of gravity, and the brain uses this information to maintain balance and orientation. In microgravity, however, the inner ear no longer senses this gravitational pull, causing confusion between what the body expects and what it experiences. This leads to symptoms such as nausea, dizziness, and headaches—symptoms similar to motion sickness on Earth, but often more severe.

About half of all astronauts experience SAS during the first few days of a space mission. While the symptoms generally subside as the body adapts to the new environment, understanding SAS is crucial, especially as humans prepare for longer missions to Mars or other planets, where astronauts will need to perform complex tasks almost immediately after arriving in space. For Polaris Dawn, monitoring how each crew member responds to microgravity will provide valuable insights into how to better prepare future astronauts for space travel.

In addition to studying the immediate effects of SAS, researchers are also interested in understanding the long-term impact of microgravity on the human body. Prolonged exposure to weightlessness can lead to muscle atrophy, bone density loss, and changes in cardiovascular function. The Polaris Dawn mission's focus on human physiology will include monitoring these changes and testing interventions that could mitigate their impact. For example, astronauts will engage in regular exercise and use resistance-based equipment to maintain muscle mass and bone density.

Studying SAS and its broader implications is also critical for the success of space tourism. As commercial spaceflight becomes more accessible to non-professional astronauts, ensuring that passengers can enjoy their journey without experiencing debilitating motion sickness will be essential. The data collected during Polaris Dawn will help inform the design of future commercial spaceflights, making space travel a more comfortable and enjoyable experience for everyone.

Space adaptation syndrome remains one of the most significant challenges for human spaceflight, but missions like Polaris Dawn offer an opportunity to better understand and address this issue. By studying how the human body responds to microgravity and developing strategies to mitigate the effects of SAS, researchers are paving the way for safer and more comfortable space missions in the future. This research is not only critical for professional astronauts but also for the future of commercial space travel, where ensuring the well-being of passengers will be a top priority.

Sarah Gillis' Musical Tribute: Art Meets Science in Space

The Polaris Dawn mission, while primarily focused on scientific exploration, is also a testament to the enduring human desire to express creativity even in the most extreme environments. One of the unique aspects of this mission is Sarah Gillis' decision to bring music into space, creating a bridge between art and science that highlights the importance of both in human culture. Gillis, an engineer and astronaut, will perform a musical tribute during the mission, demonstrating how space travel can inspire artistic expression.

Music has long played a role in space exploration, serving as a source of comfort and connection for astronauts during their missions. From NASA's use of wake-up songs to personalized playlists for space crews, music has been a constant companion for those venturing into the unknown. For Gillis, her musical tribute is a way to honor this tradition while also showcasing the role that art plays in human life, even when we are far from home.

Gillis' tribute will likely be performed in microgravity, adding a new dimension to the experience of music in space. The absence of gravity allows for unique acoustic properties, as sound waves behave differently in a weightless environment. While the basic physics of sound—vibrations traveling through air—remain the same, the microgravity environment offers new challenges and opportunities for musical performance. For example, instruments must be adapted to account for the lack of gravity, and musicians must learn to play while floating, a challenge Gillis is prepared to embrace.

This musical tribute is more than just a performance; it represents the merging of art and science, two fields that have often been viewed as separate but are, in reality, deeply intertwined. Science provides the tools and technology that make space exploration possible, while art gives meaning and context to these achievements, helping humanity understand and appreciate the significance of space travel. By bringing music into space, Gillis is highlighting the role that creativity and culture play in shaping our understanding of the universe.

Gillis' performance will also serve as a source of inspiration for future generations of scientists, engineers, and artists. As space travel becomes more accessible, it is likely that we will see an increasing number of creative endeavors taking place in space, from music and visual arts to literature and film. These artistic expressions will help humanity connect with the experience of space travel on an emotional and cultural level, ensuring that space exploration is not just a scientific pursuit but a human one.

In conclusion, Sarah Gillis' musical tribute during the Polaris Dawn mission is a powerful reminder of the importance of art in human life. By combining her expertise as an engineer with her passion for music, Gillis is demonstrating that creativity and science are not mutually exclusive but are, in fact, complementary. Her performance in space will not only be a personal achievement but also a symbol of the human spirit's capacity for wonder, innovation, and artistic expression.

The Power of Story: Reading "Kisses From Space" to St. Jude Patients

As part of the Polaris Dawn mission, a heartfelt moment will occur when the crew connects with St. Jude Children's Research Hospital patients by reading the children's book *Kisses From Space*. This initiative underscores the power of storytelling and the profound emotional connection that space exploration can foster, even for those who are not physically traveling to the stars. The book reading event is not only a way to inspire young patients at St. Jude but also a demonstration of the capacity for space travel to bring hope and joy to people around the world.

Kisses From Space is a story written with the intent of making space exploration relatable to children, particularly those facing the challenges of illness. The book tells a tale of adventure, courage, and love, themes that resonate deeply with the children at St. Jude who are undergoing treatments for serious medical conditions. By reading this story from space, the Polaris Dawn crew is offering a unique and uplifting experience for the patients, giving them a sense of wonder and possibility.

The event also serves as a reminder of the importance of outreach and education in space exploration. While space missions often focus on advancing science and technology, they also have the power to inspire and uplift people on Earth. For the children at St. Jude, the opportunity to hear a story read from space is a once-in-a-lifetime experience that can provide comfort and motivation as they navigate their health challenges.

Storytelling has always been a fundamental part of the human experience, and space exploration offers a new frontier for sharing stories that can inspire future generations. The act of reading *Kisses From Space* from orbit creates a powerful connection between the crew of Polaris Dawn and the children at St. Jude, showing that even in the vastness of space, humanity's capacity for compassion and empathy remains strong.

The reading of *Kisses From Space* to St. Jude patients is a powerful example of how space exploration can be used to inspire and uplift people, particularly children facing serious health challenges. This event highlights the importance of storytelling in connecting people across vast distances and serves as a reminder that space exploration is not just about scientific discovery but also about fostering human connections and hope. Through this simple yet profound act, the Polaris Dawn mission is demonstrating the transformative power of space travel to bring joy and inspiration to those who need it most.

Chapter 6: Returning Home: The Journey Back to Earth

The De-Orbit Burn: Preparing for Re-entry

The de-orbit burn is a critical phase of any space mission, and for the Polaris Dawn crew, it marks the beginning of their journey back to Earth after several days in orbit. This maneuver involves firing the spacecraft's engines to slow its speed enough to allow it to leave its orbit and descend toward Earth. Though the Polaris Dawn mission is a triumph of human ingenuity and technological innovation, the return to Earth is one of the most dangerous aspects of space travel, requiring precise timing, calculations, and preparation to ensure the crew's safety.

The de-orbit burn is typically planned with pinpoint accuracy to ensure the spacecraft re-enters the

Earth's atmosphere at the correct angle. A miscalculation could lead to two potentially fatal outcomes: a shallow angle could cause the spacecraft to skip off the atmosphere like a stone on water, while an overly steep angle could lead to an intense re-entry that the heat shield may not be able to withstand. Therefore, mission control works in tandem with the spacecraft's onboard systems to make sure the burn occurs at the exact moment necessary to initiate the descent.

Before the burn begins, the crew will prepare by securing themselves inside their seats and checking their equipment. This preparation includes ensuring their suits are correctly fastened and that all loose objects within the spacecraft are properly stowed. The de-orbit burn typically lasts only a few minutes, but it is one of the most crucial stages of the mission. During this phase, the spacecraft's speed is reduced by a few hundred miles per hour, which is enough to allow gravity to pull the craft back toward Earth.

The de-orbit burn is not just about slowing down but also about precision. The spacecraft must hit a specific target zone on the planet to ensure a safe landing. For Polaris Dawn, the aim will be to land in the Gulf of Mexico, where recovery teams will be waiting. This complex choreography between space and Earth operations underscores the immense planning and coordination required for a safe re-entry. Once the burn is complete, the spacecraft is on a fixed path toward Earth, with no room for error as it prepares to face the extreme heat and pressure of atmospheric re-entry.

Surviving the Heat: Protecting the Crew from 3,500°F

Re-entering the Earth's atmosphere is one of the most perilous aspects of space travel, as spacecraft face intense heat and pressure that could easily compromise the integrity of the vehicle and the safety of the crew. During re-entry, the spacecraft will encounter temperatures as high as 3,500°F (about 1,927°C), which is hot enough to melt most metals. The spacecraft relies on advanced heat shields to survive this ordeal and ensure that the astronauts inside remain protected.

The heat generated during re-entry is caused by friction between the spacecraft and the Earth's atmosphere. As the spacecraft speeds through the atmosphere at several miles per second, the air in front of it compresses, creating extreme heat. The outer shell of the Crew Dragon capsule is built with specialized materials designed to absorb and dissipate this heat, ensuring that it doesn't transfer to the interior cabin where the astronauts are seated.

SpaceX's Crew Dragon uses a heat shield made of a material known as PICA-X, an advanced version of Phenolic Impregnated Carbon Ablator (PICA), which was originally developed by NASA. This material is designed to slowly burn away in a process called ablation, carrying heat away from the spacecraft as it re-enters the atmosphere. This technology has been tested and proven effective in multiple missions, but the Polaris Dawn crew will still be keenly aware of the intense forces at play as they descend through the atmosphere.

The temperature inside the spacecraft, on the other hand, remains relatively stable thanks to insulation and life support systems that regulate the cabin's environment. For the crew, the sensation of re-entry is marked more by the forces of deceleration as the spacecraft slows down dramatically from its orbital speed. At this stage, the crew will experience several Gs of force pressing them back into their seats as the capsule makes its fiery descent.

Despite the intense heat outside, the crew is largely shielded from the immediate dangers, but there are always risks associated with re-entry. Engineers and mission planners meticulously design every aspect of the spacecraft's re-entry trajectory, testing and retesting each component of the heat shield. Still, even with the best technology, re-entry remains a high-stakes phase of any mission. As the capsule approaches the lower atmosphere and the temperatures begin to drop, the crew can start to anticipate their return to Earth and the imminent splashdown.

Splashdown in the Gulf: A Safe Return to Earth

After surviving the intense heat of re-entry, the Polaris Dawn crew's next milestone is a successful splashdown in the Gulf of Mexico. Splashdowns have been a staple of human space exploration since the early days of the Apollo missions, and while modern spacecraft are more advanced, the basic principles of landing in the ocean remain the same. The spacecraft must decelerate enough for a gentle impact with the water, after which recovery teams will retrieve the capsule and crew.

As the capsule reaches the lower atmosphere, it will deploy a series of parachutes designed to slow its descent further. First, a set of smaller drogue parachutes will deploy to stabilize the capsule, followed by the release of the main parachutes, which will drastically reduce the spacecraft's speed as it approaches the surface. The crew will experience a much more controlled descent during this phase, though they will still feel the jolts of the parachutes deploying and the splashdown itself.

Splashdown is a critical moment because it signals the final stage of the mission. However, landing in water presents its own challenges. The ocean can be unpredictable, with rough seas or high winds potentially complicating recovery efforts. That's why SpaceX coordinates closely with the U.S. Coast Guard and other agencies to ensure the landing zone is secure and that recovery teams are in place long before the capsule hits the water. Boats, helicopters, and divers will be standing by to assist the crew immediately after splashdown, ensuring a swift and safe recovery.

Once the capsule hits the water, the recovery process begins. Divers will approach the capsule to secure it and ensure it is stable. A recovery ship will then lift the capsule out of the water and onto the deck, where the astronauts will be extracted. For the crew, this marks the official end of their journey, as they exit the capsule and step onto solid ground—or, in this case, the deck of a recovery ship—for the first time since launch. Medical teams will be on hand to check on the astronauts' health, and after a brief assessment, they will be flown back to shore to reunite with their families.

What Comes Next for SpaceX and Private Space Exploration

The Polaris Dawn mission is not only a significant achievement for the crew and SpaceX, but it also represents the future of private space exploration.

With each successful mission, SpaceX is proving that private companies can take on increasingly ambitious roles in space, opening the door for more frequent and accessible space travel. What comes next for SpaceX will likely be determined by the success of the mission, but the company has already laid the groundwork for several future endeavors, including missions to the Moon, Mars, and beyond.

One of SpaceX's most ambitious goals is to develop a fully reusable spacecraft capable of carrying humans to Mars. The company's Starship, currently in development, is designed to be the largest and most powerful rocket ever built, capable of transporting both crew and cargo to deep space destinations. While the Crew Dragon capsule used for Polaris Dawn is a significant step forward, Starship represents the future of long-duration spaceflight and the possibility of establishing a permanent human presence on other planets.

Additionally, SpaceX's involvement in commercial space travel is paving the way for other private companies to enter the market. The success of Polaris Dawn could inspire further investment in space tourism, research missions, and even private space stations. As competition in the private space sector grows, costs will likely decrease, making space more accessible to people beyond professional astronauts.

Ultimately, Polaris Dawn is a stepping stone toward a new era in space exploration, one where government agencies like NASA collaborate with private companies to push the boundaries of what's possible. With future missions on the horizon, SpaceX is well-positioned to continue leading the charge in commercial space exploration, taking humanity to new heights and new frontiers.

Conclusion

Polaris Dawn is more than just another space mission; it stands as a landmark in the evolution of space travel. As the first mission under the Polaris Program, its achievements have pushed the boundaries of what private space exploration can achieve. By flying higher than any human-crewed mission since Apollo, conducting the first commercial spacewalk, and performing critical scientific experiments, Polaris Dawn has redefined what's possible in commercial spaceflight. Beyond its immediate accomplishments, the mission also represents a paradigm shift, where private enterprises like SpaceX are now capable of taking on roles traditionally reserved for governmental agencies such as NASA. The mission has also set a precedent for inclusivity in space, as it showcased a diverse crew of engineers and visionaries working together to accomplish something groundbreaking. By opening up space to more participants and showing the world that private missions can deliver high-stakes science and exploration, Polaris Dawn has laid the foundation for the future of space travel, inspiring generations to come.

The success of Polaris Dawn is just the beginning for both SpaceX and the broader commercial space industry. As SpaceX continues to refine its technology, the next missions under the Polaris Program are expected to take even greater strides. Future missions will likely focus on longer durations in orbit, testing advanced life-support systems, and preparing for more ambitious goals, such as returning to the Moon and eventually reaching Mars. The evolution of the Starship rocket, designed for deep space exploration, will play a pivotal role in future missions, aiming to make space travel more affordable and accessible. SpaceX's ongoing partnerships with NASA and other international space agencies also suggest that future missions will be collaborative, combining the expertise of the public and private sectors. With each new mission, we move closer to the long-held dream of human space exploration beyond Earth's orbit, including the possibility of establishing permanent human settlements on other planets.

Polaris Dawn has opened the door to a future where commercial space exploration is no longer a distant dream but an evolving reality. By successfully executing a privately funded and operated mission of such complexity, it has shown the world that space is not solely the domain of governments. The possibilities for the future are truly infinite: from space tourism to deep space research, commercial enterprises are poised to revolutionize how humanity interacts with the cosmos. With lower costs and advanced technologies, space could soon become accessible to more than just professional astronauts, paving the way for private citizens, scientists, and even artists to participate in space missions. Commercial space exploration will also accelerate the development of technologies critical to sustaining life in space, from advanced habitats to propulsion systems. As companies like SpaceX continue to innovate, the horizon of what's possible in space will continue to expand, ultimately transforming our relationship with the universe and setting the stage for humans to become an interplanetary species.